Module 5 Communicatic homeostasis and energy

Topic 1 Communication and homeostasis

Cell signalling

Complex organisms that are composed of cells, tissues and specialised organs must be able to communicate with all their internal tissues and structures as well as with the external environment. One aspect of communication is the detection of changes or stimuli, and the transfer of information to allow a response to those changes or stimuli. Communication in animals is by nerve impulses and neurosecretions in the nervous system, and by chemical signals from the endocrine system. These are examples of cell signalling. Plants have fewer methods of communication.

1 **Complete the table below to compare cell signalling by nervous or hormonal communication. (AO1, AO2)** 6 marks

Feature	Signalling by nervous communication	Signalling by hormonal communication
Nature of the signal		
Method of transmission		
Target		
Time period		
Method of removal of communication signal		
Type of structures involved or required		

2 a **There is a difference between paracrine and endocrine cell signalling. Describe this difference. (AO1)** 1 mark

....

....

b **Explain the basis on which these two signalling pathways are categorised. (AO1)** 2 marks

....

....

....

c **An exocrine gland is different from an endocrine gland. Explain how these two glands are different from each other. (AO1)** 1 mark

....

3 **Referring to the diagrams, give two examples of cell signalling molecules in animals. For each example you have given state its role. (AO1, AO2)** 4 marks

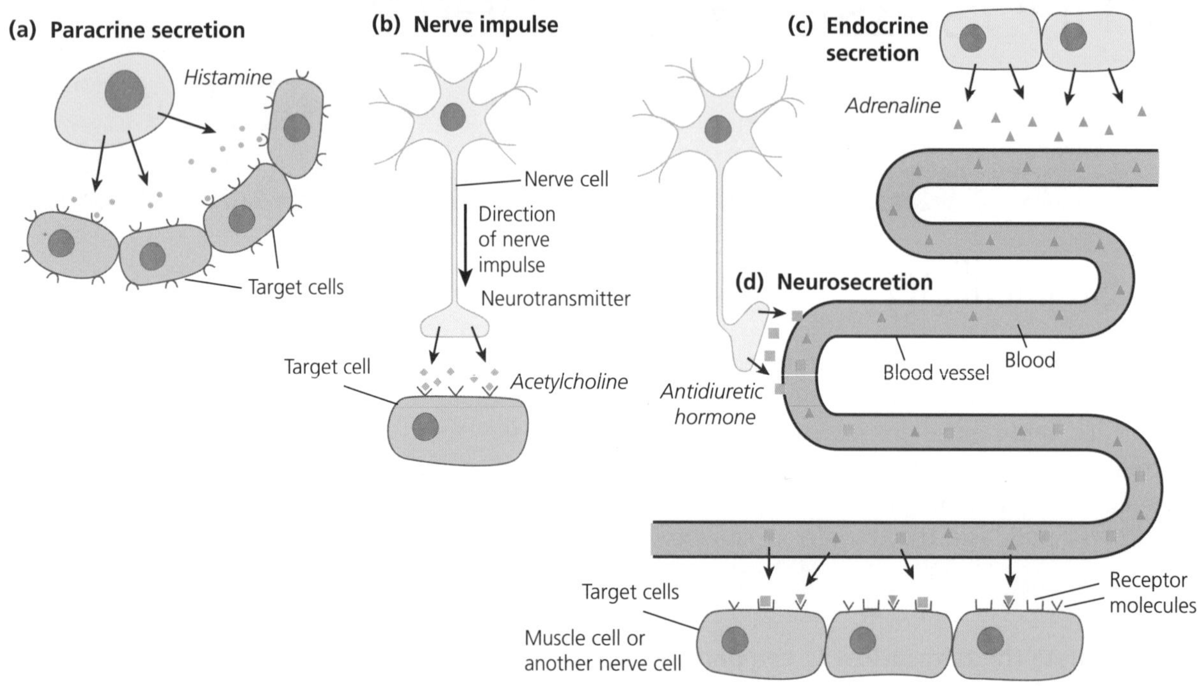

....

....

....

Homeostasis

All organisms need to maintain a constant internal environment to enable all life processes to proceed. This may be achieved either by internal balancing processes or by behaviour of the organism. Ensuring that the internal environment is kept constant is called homeostasis. Homeostasis requires that information about the conditions within and around the organism is detected.

1 **Define the word homeostasis. (AO1)** 1 mark

....

....

2 **There are four key principles that allow a constant internal state in the body to be maintained (homeostasis). The ability to detect changes is one of these principles. What are the other three? (AO1)** 2 marks

....

....

....

....

3 Complete the following table with the heading for each column, and either ticks or crosses to compare two of the main mechanisms that are responsible for maintaining a constant internal environment. One row has been completed for you. (AO1, AO2) **5 marks**

Feature of mechanism	____________ feedback	____________ feedback
Receptor detects change		
Change initiates a series of events		
Hormones activated		
A reversal of change	✓	✗
An increase in the event		

4 Describe and explain *one* example of each of the two types of mechanism in question 3. (AO1, AO2) **4 marks**

....

....

....

....

....

5 Complete the following flow diagram that summarises a generalised negative feedback mechanism. (AO1) **3 marks**

____________ detect change in conditions

↓

The ____________ receives the information: sends ____________ instructions

↓

The ____________ bring about the corrective actions as instructed by the ____________

↓

The ____________ is a return to the set point

Thermoregulation: control of body temperature

The control of body temperature is known as thermoregulation. Mammals and birds are endotherms, meaning that they have physiological mechanisms to control their body temperature. All other members of the animal kingdom are ectotherms and they have to rely on behavioural adaptations to ensure that their core temperature does not sink too low or rise too high.

1 Complete the following passage on temperature control. (AO1) **5 marks**

____________ control their internal temperature by ____________ heat and retaining it within the body when cold, or by ____________ heat when they are hot.

____________ gain and lose heat from the environment using behavioural mechanisms.

This may mean that they are ________ in the early morning until enough heat is gained, which makes them vulnerable to ________ . However, many ________ keep their internal temperature relatively ________ by their actions, so that it may only fluctuate by a few degrees. A lizard, for example, moves into the shade when it is too ________ and basks in the sun when too ________ .

2 **The optimum temperature for human enzymes is about body temperature, 37°C, and there is little fluctuation either side. What is the effect of significant variation from this optimum on the rate of enzyme-controlled reactions? (AO1)** 1 mark

....

....

3 **In thermoregulation there are control mechanisms to ensure that the internal temperature remains stable. The flow diagram shows the changes that occur in endotherms when temperature fluctuates.**

Complete the flow diagram to summarise temperature control in endotherms. (AO1) 9 marks

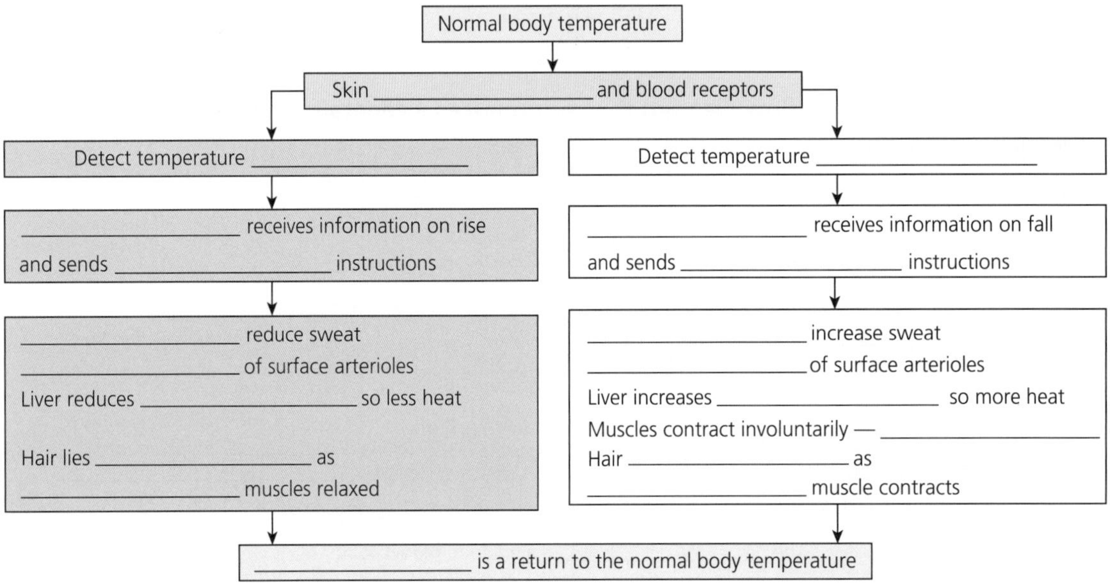

4 **Explain each of the following processes: (AO1, AO2)**

a **vasoconstriction** 2 marks

....

....

b **vasodilation** 2 marks

....

....

c **sweating** 2 marks

....

....

d shivering 2 marks

5 There are two main hormones involved in thermoregulation. Name these two hormones and outline their roles in thermoregulation. (AO1, AO2) 4 marks

6 Suggest how the liver is involved in temperature regulation (AO2) 2 marks

7 Suggest why any homeostatic process requires two control mechanisms. (AO2) 2 marks

8 How is temperature monitored in the body of an endotherm? (AO2) 2 marks

9 A teacher carried out an experiment on temperature regulation in a group of five students. The skin temperature of each student was monitored using a thermocouple attached to the forefinger of the left hand. Their internal temperature was monitored using an ear thermometer (a thermocouple inserted into the ear canal). Each student's right hand was then placed into a bucket of iced water and both the skin and internal temperatures were monitored over a period of 12 minutes, unless the student requested a halt to the experiment. The results were recorded in the table.

Time/s	Mean internal temperature/°C	Skin temperature of each student/°C	Mean skin temperature/°C
0	37.0	31.0, 31.5, 30.5, 31.0, 30.5	
2	36.8	31.0, 31.0, 30.5, 30.0, 30.5	
4	36.5	31.0, 31.0, 30.0, 29.0, 30.5	
6	36.6	30.0, 30.0, 30.0, 29.0, 30.0	
8	36.7	30.0, 30.0, 29.5, 29.0, 29.5	
10	37.2	29.5, 29.5, 29.0, 29.5, 29.0	
12	37.1	29.0, 28.5, 29.0, 29.0, 28.0	

a Calculate the mean skin temperature at each time interval. (AO2, AO3) **5 marks**

b What do the mean results suggest about internal body temperature? (AO2, AO3) **2 marks**

c Describe what happened to the mean body temperature during the 12 minutes. (AO3) **2 marks**

d What do these results suggest about temperature regulation and the influence of skin temperature on the control of internal temperature? (AO3) **2 marks**

26

Exam-style questions

1 One feature of homeostasis is osmoregulation. This maintains the water potential of the blood within certain set limits. The hormone ADH (anti-diuretic hormone) is responsible for achieving this balance.

a If the water potential of the blood plasma is allowed to increase to a significant level, what effect might this have on red blood cells? Explain your answer. **3 marks**

b ADH does not remain in the blood permanently and must be removed once it has carried out its role in osmoregulation. Suggest where ADH is removed from the blood and describe what will happen to the molecule of ADH once it is removed. **3 marks** c ADH is made up of a polymer chain, one end of which is an amine group.

i What bonds will hold this chain together? **1 mark**

....

ii What type of monomer makes up this chain? **1 mark**

....

2 a Explain what is meant by homeostasis. **2 marks**

....

....

....

b Describe how internal temperature is regulated when an endothermic animal is subjected to low external temperatures. **4 marks**

....

....

....

....

....

....

3 Explain the homeostatic control of water balance when an animal has a reduced supply of water. You should use appropriate technical terms, correctly spelt, and answer in the correct sequence. **5 marks**

....

....

....

....

....

....

....

....

Topic 2 Excretion as an example of homeostatic control

Excretion

Animals have excretory systems that allow the removal of waste products produced during the body's metabolic processes. If these waste products were allowed to accumulate, they would have toxic effects that would damage cells.

1 **What is the definition of excretion? (AO1)** 1 mark

....

....

2 **Complete the table to:**

a **Give the process creating the waste production in each case. (AO1, AO2)**

b **For each, state where this takes place. (AO1)**

c **Explain the effect on the body if each waste product is allowed to accumulate and is not excreted. (AO2)** 6 marks

Metabolic waste product	Process creating waste product	Location in the body	Effect on the body if not excreted
Carbon dioxide			
Water			
Urea			
Bile pigments			
Uric acid			
Ammonia			

3 Explain how the lungs, liver and kidneys contribute to homeostasis in the body. (AO2) 1 mark

4 Water is a waste product of metabolism, but water is not usually excreted in large quantities. Suggest why. (AO2) 2 marks

The liver

1 The liver is made up of several lobes, each of which is subdivided into lobules that are separated by connective tissue. Within each lobule there are numerous rows of cells called hepatocytes. Explain why the lobule is called the functional unit of the liver. (AO2) 2 marks

2 State the adaptive features of hepatocytes and explain the advantages to the cells of each adaptation. (AO1, AO2) 4 marks

3 The diagram shows a small section of a liver lobule with the hepatocytes arranged in radiating rows. Add arrows to the diagram to show the direction of blood flow. Label the hepatocytes, the bile canaliculus and the three blood vessels shown. (AO1) 7 marks

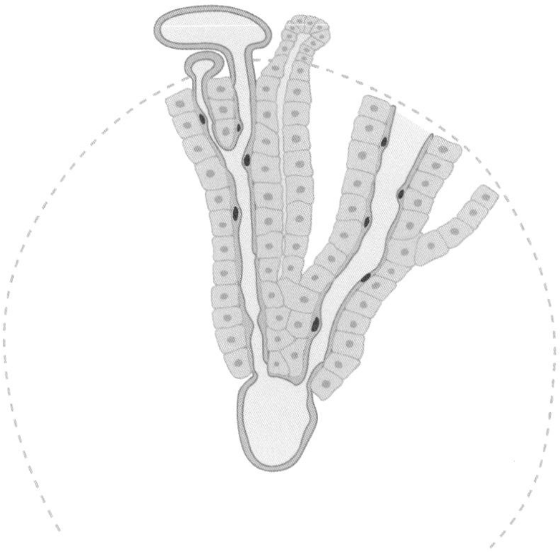

4 **The table lists some of the many functions of the liver. Complete the table to outline each process associated with that function, and name the products. (AO1)** 12 marks

Liver function	Process	Product
Formation of excretory products		
Storage of carbohydrate		
Detoxification of hormones, alcohol, drugs		
Lactate production and metabolism		
Bile formation		
Breakdown of hydrogen peroxide		

5 **Liver cells manufacture the blood plasma protein albumin, which is important in maintaining the water potential of the blood plasma. Suggest why a severe lack of protein in the diet or liver failure may result in an accumulation of fluid in the abdominal cavity. (AO2)** 2 marks

....

....

....

6 **There are a number of coenzymes involved in metabolic pathways.**

a **Give *two* examples of a coenzyme. (AO1)** 1 mark

....

b **Explain the role of one of the coenzymes in metabolism. (AO2)** 2 marks

....

....

....

7 **Name three metabolic pathways that are cyclical. (AO1)** 3 marks

....

....

....

The kidney

The kidneys are the main excretory organs situated at the top back of the abdomen just below the diaphragm. The kidneys filter blood and excrete waste products and substances in excess. They are also the effector organ in the regulation of water in the body.

1 **The kidney is part of the urinary system. Name the other structures that form part of this system. (AO1)** 1 mark

....

....

2 **Name the parts labelled A–I in the diagrams below that show the gross structure of the kidney. (AO1)** 7 marks

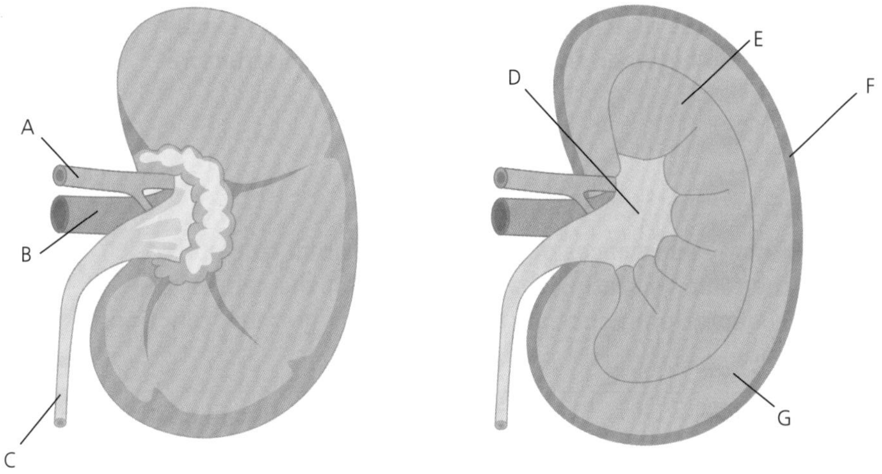

A .... **B** .... **C** ....

D .... **E** .... **F** ....

G ....

3 **The function of the kidney and the formation of urine are closely linked and are considered in parallel. For each of the kidney structures, state the function and the adaptive features involved in these processes. (AO1)** 14 marks

a **Bowman's capsule**

....

....

b **glomerulus**

....

....

c **proximal convoluted tubule**

....

....

d **loop of Henlé**

....

....

e **distal convoluted tubule**

....

....

f **collecting duct**

....

....

g **ureter**

....

....

4 **The basement membrane of the Bowman's capsule acts like a sieve for molecules passing from the capillary into the filtrate. It only permits substances that have a relative molecular mass (RMM) of less than 69 000 to be filtered through. Explain the significance of this relationship between the basement membrane structure and RMM. (AO2)** 2 marks

....

....

5 a **Selective reabsorption of many substances occurs in the proximal convoluted tubule (PCT). Complete the table by stating the adaptations of the cells of the PCT and relate how each adaptation is specialised to allow the greatest functional efficiency. (AO2)** 6 marks

Structure of PCT	Adaptation	Specialisation in relation to function
Microvilli	Increases and provides a large surface area	Increasing surface area increases possible number of symport proteins
Symport proteins		
Mitochondria		
Na^+/K^+ pump		
Tight junctions between the cells		

b What happens to the volume of the filtrate as it leaves the proximal convoluted tubule? (AO1) **1 marks**

6 The remainder of the nephron regulates blood concentration. Suggest *two* different control methods that may achieve this regulation. (AO2) **2 marks**

Regulation of water content in the blood

Water regulation occurs in the loop of Henlé, the surrounding tissues and the collecting duct.

1 Give a detailed explanation of the countercurrent exchange mechanism in the medulla of the kidney. (AO1, AO2) **5 marks**

2 There are two types of loops of Henlé: long loops that extend into the medulla and short loops that do not extend into the medulla. Human kidneys have both types of loop.

a Suggest which type of loop of Henlé is most beneficial in conserving water from the filtrate before it becomes urine. (AO2) **2 marks**

b What is the mechanism that allows this to occur. (AO1) **2 marks**

3 a Name the hormone that is involved in the control of ion balance in the blood, and state from where it is secreted. (AO1, AO2) **2 marks**

b How does this hormone control ion balance? (AO1, AO2) **2 marks**

....

....

....

4 a Which hormone is involved with regulating blood water balance? (AO1) **1 mark**

....

b There is an unusual aspect to the production and release of this hormone. Describe what this is. (AO1, AO2) **2 marks**

....

....

....

....

c Describe the mechanism that is used to control blood water balance. (AO1, AO2) **3 marks**

....

....

....

....

Kidney failure

Kidney failure occurs for a variety of reasons. It may be permanent or may be temporary. In all cases, the result is a build-up of toxins, urea, salts and water because they cannot be excreted. It will quickly be fatal. The treatment is either dialysis or a transplant.

5 a What is dialysis? (AO1) **1 mark**

....

b Suggest the expected composition of dialysis fluid. (AO1, AO2) **2 marks**

....

....

....

6 Explain the differences between the two types of dialysis. (AO1, AO2) **4 marks**

....

....

....

....

....

....

....

Testing urine

7 Complete the table to list some different types of urine test and describe the condition that each is testing for. (AO1, AO2) **6 marks**

Test	Condition being tested for
Glucose	
Glucose tolerance	
	Indicates high blood pressure or kidney infection or malfunction
	Indicates bacterial infection
	Pregnancy test as this hormone is only produced during pregnancy
Steroids	

Exam-style questions

1 *Notomys alexis*, the Australian hopping mouse, has a great facility for water retention as it lives in desert conditions. A similar animal, the Mongolian gerbil, *Meriones unguiculatus*, has become a very popular pet because it produces only a small amount of urine. Suggest three adaptations that may reduce water loss and encourage water retention in desert animals such as the hopping mouse and the gerbil. **2 marks**

....

....

....

2 Complete the following passage that describes the structure of the kidney, by writing the correct word in each space. **6 marks**

Each kidney is made up of millions of tubules called ________________. These do not lie in one plane, but instead twist and are irregular in shape. Each ________________ filters the blood plasma and produces a filtrate, which consists of waste substances, some useful substances and water. Some of the useful substances and quantities of the water are reabsorbed before the filtrate is emptied into the ________________ ________________. The tube itself begins with a cup-shaped structure called the ________________ ________________, in which sits a tightly grouped knot of capillaries called the ________________. These are found in the ________________ region. From here the tubule twists and convolutes to form the ________________ ________________ ________________.

The ________________ ________________ ________________ is a hairpin bend in the medulla region, and the ________________ ________________ ________________ is also twisted. At this point the tubule connects with the collecting duct, which empties into the ________________ ________________, the ________________ and finally into the bladder.

3 a **Explain what happens as the filtrate moves down the descending loop of Henlé.** **3 marks**

....

....

....

....

....

b **Explain how the conditions in the loop of Henlé favour the movement of water out into the tissues by osmosis.** **4 marks**

....

....

....

....

....

4 **Define the following words:**

a **aquaporin** **1 mark**

....

b **antiport** **1 mark**

....

Topic 3 Neuronal communication

A neurone is a nerve cell. These cells make up the nervous system, that is, the brain, the spinal cord and all the peripheral nerves. Each nerve comprises a bundle of neurones. The nervous system, and thus the individual neurones, represents one of the key systems that enables communication within and between body parts, and so instructs the coordination of body activities.

Neurones

1 **How do neurones connect with other neurones? (AO1)** **1 mark**

....

....

2 **Most structures in the body are able to repair damage, but this is not generally possible for nerves. Explain why nerves do not generally repair themselves. (AO1)** **2 marks**

....

....

....

3 **Complete the table to summarise the differences between the three different types of neurone. (AO1, AO2)** 7 marks

Feature	Motor neurone	Sensory neurone	Relay neurone/interneuron
Position of cell body			
Presence of dendrites on cell body			
Axons			
Axon terminals			
Myelin sheath			
Neurone connections			
Direction of impulse			

Receptors

Receptors receive information from either the internal or the external environment and convert that information into another form of energy.

1 **What process occurs in structures like receptors that are able to convert energy from one form to another? (AO1)** 1 mark

....

2 **Choose *three* types of human receptor and state the role of each in the human body. One example is given for you. (AO1, AO2)** 3 marks

Receptor	Role
Photoreceptor	Detects light

Nerve impulse

Nerve cells work by generating and transmitting an electrical current called a nerve impulse.

Resting potential

1 **Give a detailed explanation of how the resting potential of a neurone is established. (AO2)** 4 marks

....

....

....

....

....

2 **What role does active transport play in establishing the resting potential of the neurone? (AO1, AO2)** 2 marks

....

....

Action potential

3 **An impulse is generated when the resting potential is converted into an action potential. Outline the steps involved in this process. (AO2)** 5 marks

....

....

....

....

....

....

....

4 **What is the role of voltage-gated channels in the transmission of a nerve impulse? (AO2)** 4 marks

....

....

....

....

....

....

5 **The events that occur when an impulse is generated from converting a resting potential into an action potential are detailed below. Complete the information given by filling in the blanks. (AO1)** 5 marks

At resting potential the voltage-gated channels are closed, preventing ______________ ions entering the axon. A stimulus causes the ______________ ______________ ion channels to open and the concentration gradient of ______________ ions and ______________ gradient both allow sodium ions to enter the cells by ______________. This raises the resting potential of the membrane from ______________. If the potential of the membrane reaches ______________ all ______________ channels open and the ions flood into the axon.

6 **The establishment of an action potential is an example of positive feedback. Explain what this means in the context of the neurone membrane. (AO2)** 2 marks

....

....

7 **What is the threshold potential and what is its significance in nerve impulse transmission. (AO1, AO2)** 2 marks

....

....

....

....

8 **Complete the table below to explain the difference between depolarisation, repolarisation and hyperpolarisation. The resting potential has been completed for you. (AO1, AO2)** 3 marks

Event	Resting potential	Depolarisation	Repolarisation	Hyperpolarisation
Membrane potential	−70 mV			
Voltage-gated sodium channels	Closed			
Voltage-gated potassium channels	Closed but membrane leaky			
Ion balance inside	High K^+ Low Na^+			
Ion balance outside	Low K^+ High Na^+			

9 a **Use the graph below to explain why a second stimulus cannot trigger a second impulse until a time period of about 2 ms has passed since the previous impulse. (AO2)** 2 marks

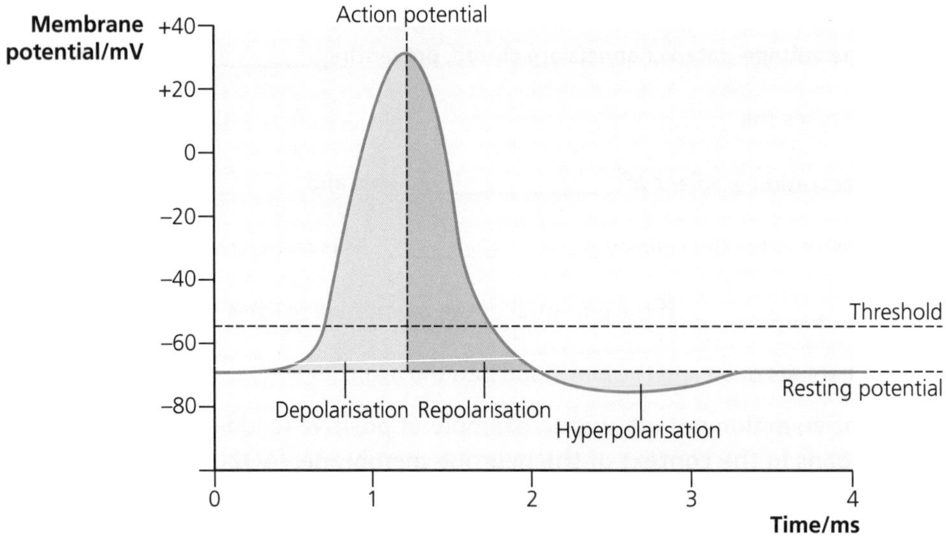

....

....

....

b **Explain the significance of this time period. (AO2)** 2 marks

....

....

10 **How is the action potential propagated along the axon? (AO1, AO2)** 3 marks

....

....

....

....

The all-or-nothing principle

11 **What is the all-or-nothing principle? (AO1)** 2 marks

....

....

....

12 **How does the strength of an impulse vary if a bigger stimulus cannot produce a bigger impulse? (AO1, AO2)** 2 marks

....

....

....

The myelin sheath

13 What are the properties of the myelin sheath and what role does it play? (AO1, AO2) **4 marks**

....

....

....

....

14 Explain how saltatory conduction increases the speed of conduction in a myelinated neurone, when compared with conduction in a non-myelinated neurone. (AO2) **3 marks**

....

....

....

....

....

Synapses

15 The following diagram of a synapse shows the structures involved with the transmission of an impulse. Name the structures labelled A–H. (AO1) **3 marks**

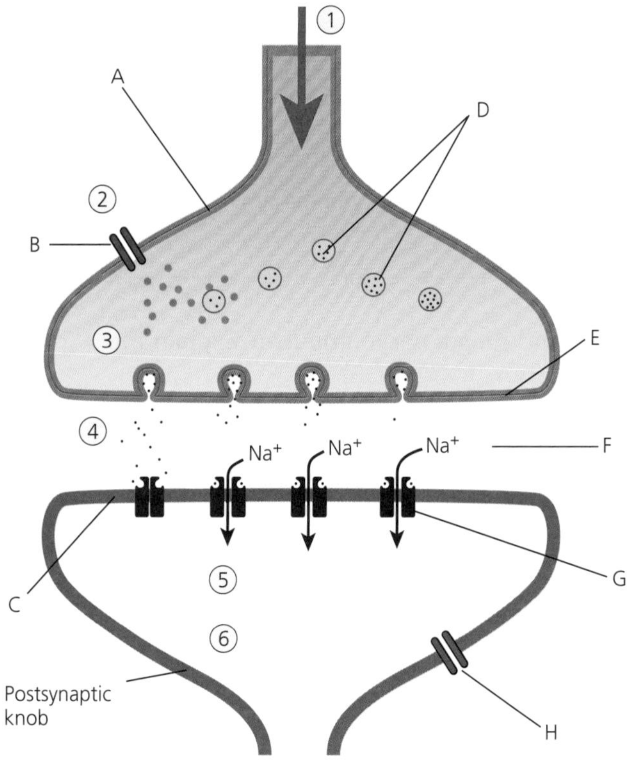

A ....

B ....

C ....

D ....

E ....

F ....

G ....

H ....

16 The stages involved in transmission of the impulse across the synaptic gap are labelled 1–6 in the above diagram. Outline below what takes place at each stage. (AO1, AO2) **6 marks**

....

....

....

....

....

....

....

....

....

17 What is the role of cholinesterase? (AO1, AO2) **3 marks**

....

....

....

....

18 Complete the table to compare excitatory and inhibitory synapses. (AO1, AO2) **4 marks**

	Excitatory synapses	Inhibitory synapses
Action potential		
Ion channels		
Membrane potential		
Example		___________ (gamma-aminobutyric acid)

19 How does summation improve the chances of initiating a nerve impulse? Explain the difference between temporal and spatial summation. (AO1, AO2) **5 marks**

....

....

....

....

....

....

....

....

20

Exam-style questions

1 Dendrites are the parts of the motor neurone that transmit impulses:

A towards the cell body

B away from the cell body

C to the central nervous system

D from the node of Ranvier **1 mark**

2 The cell body of a sensory neurone is found:

A at one end of the neurone in the central nervous system

B in the middle of the neurone axon

C to one side of the neurone in the dorsal root ganglion

D at the receptor end of the neurone **1 mark**

3 The best explanation of a nerve impulse is:

A a self-propagating change in polarity across a membrane

B a sudden depolarisation of the nerve cell membrane

C the active removal of sodium ions across the membrane

D the spread of a localised wave of electrons down the axon **1 mark**

4 Nerves form one type of cell signalling in the body.

a Explain what is meant by cell signalling. **2 marks**

....

....

....

b What system other than the nervous system is involved in cell signalling? **1 mark**

....

5 A stimulator delivered an electrical shock of 0.5 V to the axon of a nerve fibre and the response was detected and recorded by a microelectrode, in millivolts, at a point 10 cm further along the axon. A graph was plotted of the membrane potential in mV against time in milliseconds.

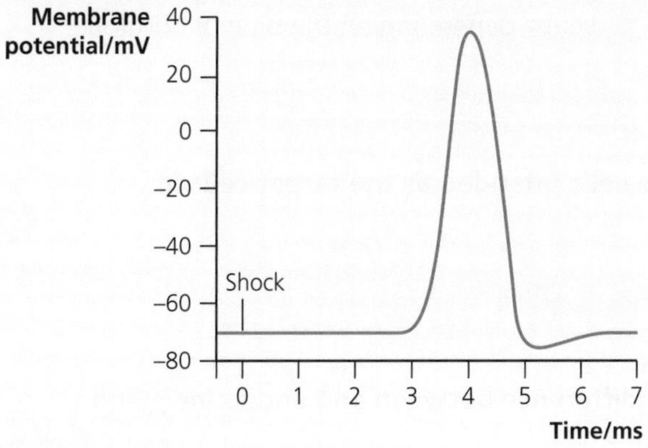

a **What is the resting potential of the nerve fibre as recorded in the graph?** 1 mark

....

b **Why is there a delay between delivery of the shock and the start of the response?** 1 mark

....

....

c **Given that the distance between the stimulus and detection of the response is 10 cm, calculate the conduction velocity of the nerve fibre in $m\,s^{-1}$. Show your working.** 3 marks

d **If a larger shock of 1.0 V was delivered instead of the 0.5 V, what response would you expect?** 2 marks

....

....

....

....

e **Draw a line on the graph to represent the response that you would expect if a shock of 0.05 V was delivered.** 1 mark

Topic 4 Hormonal communication

Hormones are chemical messengers that play an important role in the control of body functions. Hormones act together with the nervous system, but unlike the nervous system they are not limited to controlling those structures to which they are attached. Hormones travel around the body in the blood, and so can reach and control tissues and organs in different parts of the body. They are also able to exert their control over long periods of time, which can range from minutes to many years, depending on the type of hormone.

1 **How are hormones only able to control those cells intended as the target cells in the body? (AO1)** 1 mark

....

....

2 **Giving an example of each type, describe the difference between and endocrine gland and an exocrine gland. (AO1)** 2 marks

....

....

....

Adrenaline

3 **Adrenaline is produced in the medulla of the adrenal glands. What is the main function of this hormone? (AO1)** 1 mark

....

4 a **Why can adrenaline not pass directly through the cell surface membrane of a target cell? (AO2)** 2 marks

....

....

b **Give a detailed description of how the primary and secondary messengers act to exert their effect on the target tissue. (AO2)** 2 marks

....

....

....

5 **Give *two* examples of secondary messengers. (AO1)** 1 mark

....

6 **Complete the flow diagram below to show the steps that occur between adrenaline reaching a target cell and exerting its influence on the cell. (AO1, AO2)** 5 marks

Adrenaline binds to an adrenergic receptor on the target cell surface membrane

↓

______________ complex activates ______________

↓

______________ converted to ______________

↓

______________ activates the enzyme ______________

↓

______________ catalyses ATP to ______________

↓

______________ sets up a cascade of reactions within the cell

The pancreas

7 The pancreas is located under the stomach and next to the duodenum. It contains many exocrine cells and also small patches of tissue called the islets of Langerhans. Describe the substances that these the two types of cell in the pancreas secrete. (AO1) 2 marks

....

....

....

The control of blood glucose concentration

8 The hormones produced by the pancreas are important in the control of the blood glucose concentration. Why is it important that the concentration of glucose in the blood is maintained at between 4 mmol dm^{-3} and 8 mmol dm^{-3}? (AO2) 3 marks

....

....

....

....

9 Define the following words: (AO1)

a glycogenolysis 1 mark

....

b gluconeogenesis 2 marks

....

....

c glycogenesis 1 mark

....

d glucagon 2 marks

....

....

e glycogen 1 mark

....

Insulin and glucagon

10 Outline the correct sequence of events that lead to the release of insulin. (AO1) 3 marks

....

....

....

⑪ Where is glucagon produced and what function does this hormone perform? (AO1) 3 marks

....

....

....

⑫ Complete the table to compare the effects of insulin and glucagon on the function of their target cells and tissues.

Feature	Effect of insulin	Effect of glucagon
Receptors		
Permeability of cell surface membranes		
Enzyme action		
Effect on cell respiration		
Molecular conversion		
Effect on blood glucose concentration		
Effect on the cells of the Islets of Langerhans		

⑬ Both insulin and glucagon use a negative feedback mechanism to exert control of blood glucose levels. Explain what is meant by negative feedback control? (AO1) 1 mark

....

....

⑭ Complete the flow diagram to show how insulin and glucagon control blood glucose concentration. (AO1, AO2) 4 marks

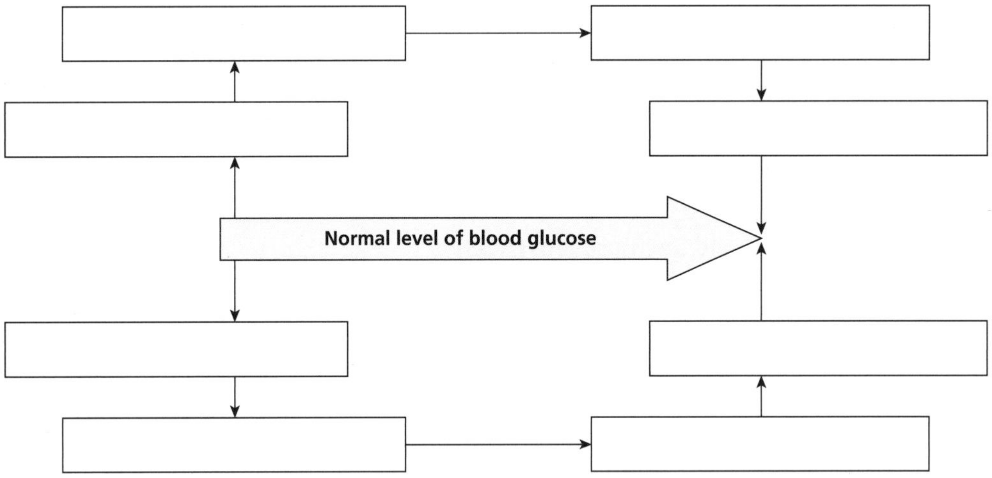

15 Diabetes is an example of a chronic condition. What is a chronic condition? (AO1) 1 mark

....

16 Referring to the action of the pancreas, the immune system and genetic factors, compare the two types of diabetes and state the treatment for each. (AO1, AO2) 5 marks

....

....

....

....

....

17 Recent research has suggested that cinnamon may improve type 2 diabetes. It was decided to test this suggestion in 100 type 2 diabetic patients. The 100 patients were subdivided into four groups, and each group was given a different cinnamon regime over a period of 40 days.

Each patient's blood glucose concentration was tested before the treatment and again at the end, and the mean blood glucose concentration for each group was recorded. The mean percentage change in blood glucose concentration for each group was then calculated. The results are shown in the table.

Daily treatment	Mean blood glucose concentration at start of regime/mmol dm^{-3}	Mean blood glucose concentration at end of regime/mmol dm^{-3}	Mean percentage change in blood glucose concentration/ mmol dm^{-3}
1 g cinnamon	12	9	
3 g cinnamon	13	10.66	
1 g placebo	11.5	11.73	
3 g placebo	12.5	12.75	

a Calculate the mean percentage change in blood glucose for each treatment regime. (AO3) 4 marks

b What error has been made when recording these data? (AO3) 1 mark

....

....

....

c The investigators' method of recording the data for each group as a mean was criticised by a peer review. Why may recording only the mean value lead to misleading results? (AO3) 2 marks

....

....

d Suggest how this method could be improved. (AO3) 1 mark

....

....

18 **Insulin has historically been obtained from the pancreatic tissue of animals such as cattle or pigs. It is now produced using genetically modified bacteria. What are the *advantages* of human insulin produced in GM bacteria? (AO2)** 3 marks

....

....

....

50

Exam-style questions

1 **Put the following sentences, A–J, into the correct sequence.** 5 marks

A **Cyclic AMP (cAMP) activates an enzyme.**

B **Calcium ions flood into the cell.**

C **ATP is converted to cAMP.**

D **ATP causes potassium channels to close.**

E **Glucose is respired and releases ATP.**

F **Glucose is converted to glycogen.**

G **Insulin is released.**

H **The enzyme adenyl cyclase catalyses the reaction.**

I **Glucose levels rise and glucose moves into beta cells.**

J **Insulin binds to receptors.**

....

2

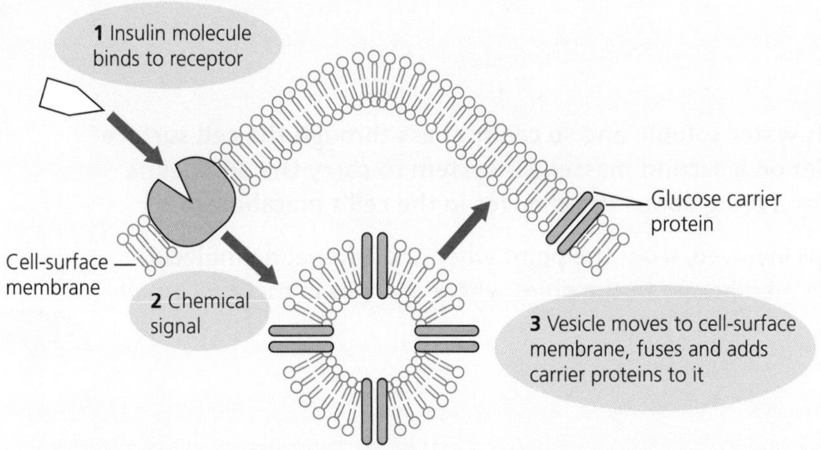

a **Referring to the diagram, explain why insulin can only have an effect on some cells and not on others.** 1 mark

....

....

b **What is chemical signal 2?** 1 mark

....

....

c How would the mechanism shown in the diagram affect the uptake of glucose by the cell? **2 marks**

d Give *two* examples of how insulin can affect blood glucose levels. **2 marks**

e What is diabetes? **2 marks**

f If a diabetic patient is maintained on a combination of a controlled diet and insulin injections how might exercise affect their glucose balance? **2 marks**

g What is RQ? **1 mark**

h Explain why the RQ of an untreated diabetic patient may be lower than that of a non-diabetic. **3 marks**

3 The adrenaline molecule is water soluble and so cannot pass through the cell surface membrane. Instead, it relies on a second messenger system to carry the adrenaline signal into the target cell and bring about the changes in the cell's metabolism.

a Give details of the steps involved, from the point when the adrenaline molecule reaches the cell surface membrane, to the point where changes in metabolism of the cell occur. **4 marks**

b The target cell has a multi-stage second messenger system within the cytoplasm, which amplifies the effects that just a small number of adrenaline molecules can have on the cell's metabolism. Suggest how the many steps in the signalling pathway allow only a few adrenaline molecules to rapidly cause very large effects. **2 marks**

....

....

c Where in the body is adrenaline produced? **1 mark**

....

4 **Secondary messengers are needed when the primary messenger cannot pass through the lipid-based cell membrane and enter the cell. Suggest how the following substances act as second messengers:**

a **calcium ions** **3 marks**

....

....

....

b **IP_3 (inositol-1,4,5-triphosphate)** **3 marks**

....

....

....

5 **The chemical composition of hormones allows them to be subdivided into three groups. Give an example of each group and describe the difference in composition between them.** **3 marks**

....

....

....

....

Topic 5 Plant responses

When organisms respond to changes in the environment they need a receptor to detect that change (the stimulus), as well as a means of communicating the change, and an effector to bring about a reaction to the change. Most plant responses to the environment are by growth responses called tropisms, which are controlled by hormones. Plant hormones are often called plant growth regulators (PGRs).

Plant growth regulators

1 Name five plant hormones and, for each hormone, give the reaction that it initiates. (AO1, AO2) 5 marks

....

....

....

....

....

2 There are a number of different plant responses that all confer an advantage to the organism. Explain the difference between the following paired responses. (AO1, AO2) 4 marks

a **tropism and a nastic response**

....

....

b **phototropism and geotropism**

....

....

c **thigmotropism and chemotropism**

....

....

d **the response to predation by herbivores (herbivory), and the response to abiotic stress**

....

....

Phototropism

3 Using only two or three sentences, summarise the experiments on phototropism that were carried out by each of the following researchers. (AO2) 8 marks

a **Darwin**

....

....

....

b **Boysen-Jensen**

....

....

....

c **Paál**

....

....

....

d **Went**

....

....

....

4 **Indole-3-acetic acid (IAA) is an example of an auxin. This functions by controlling how genes are expressed. How does IAA bring about the response seen in phototropism? (AO1, AO2)** 4 marks

....

....

....

....

....

Geotropism

5 **Explain how the geotropic response by a plant root to auxin is different from the phototropic response by a shoot to auxin. (AO1, AO2)** 3 marks

....

....

....

....

6 **Gibberellin is another plant hormone that stimulates growth. Gibberellin works with auxin in a synergistic way. Explain the term synergistic. (AO2)** 1 mark

....

....

Leaf loss

Plants lose their leaves in the autumn, unless they are evergreen. Those plants that do lose their leaves are called deciduous plants. Leaf loss has a number of advantages for the plant and is controlled by hormones.

7 What are the advantages to plants of losing their leaves before the onset of winter? (AO1) **3 marks**

....

....

8 Complete the flow diagram to show the steps involved in the control of leaf loss, and provide details of the role played by the hormones that are involved. (AO1, AO2) **3 marks**

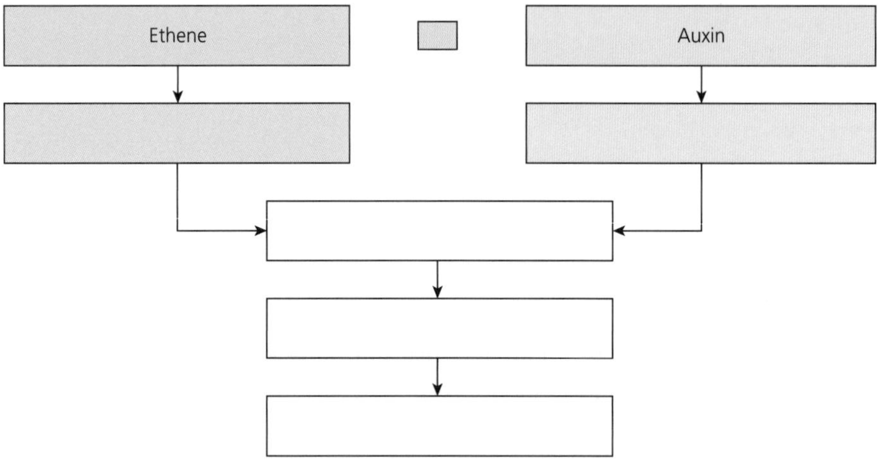

9 The following table lists the stages and hormones involved in stomatal closure. Complete the second column to explain the role that each one plays. (AO2) **2 marks**

Chemical/process	Role in stomatal closure
ABA (abscisic acid)	
Calcium channels	
K^+, NO_3^- and Cl^- ions	
Osmotic potential of cell	
Turgor of guard cells	

10 What is the role that gibberellins play in seed germination? (AO2) **2 marks**

....

....

Commercial use of plant hormones

11 For each plant hormone below give details of how it is used to control growth and development in commercially grown plants. (AO2) **8 marks**

a **auxin**

....

b **ethene**

....

c **gibberellin**

....

d **cytokinin**

....

Exam-style questions

1 **Abscisic acid (ABA) is synthesised in leaves and affects the stomata. An investigation into the effect of ABA on stomata was carried out on a healthy maize plant that was not watered for 5 days. The plant was then watered again on day 6. Fully open stomata have only a low resistance to air flow.**

Three different measurements were taken over the period of the investigation:

1 **water potential of the leaves**

2 **the resistance to the flow of air through the stomatal opening**

3 **the ABA content of the leaves**

The results are shown in the table.

Day	1	2	3	4	5	6	7	8	9	10
Water potential of leaves/mPa	−0.64	−0.40	−0.32	−1.12	−1.32	−1.44	−0.24	−0.48	−0.48	−0.56
Air flow resistance/ arbitrary units	4	4	4	20	70	68	35	16	12	6
ABA content of leaves/arbitrary units	0.10	0.10	0.10	1.85	1.9	2.15	0.90	0.10	0.10	0.10

a **What type of compound is ABA?** 1 mark

b **Comment on the change in water potential between days 3 and 6.** 1 mark

c **Explain the likely cause of this difference.** 3 marks

d **A hypothesis was suggested: 'ABA is synthesised during times of water stress to cause the stomata to close.' Use the data to comment on this hypothesis.** 2 marks

2 **Plant responses to any environmental change are coordinated by plant growth substances.**

a **Why do plants need to respond to their environment?** 2 marks b Scientists investigated the effects of plant growth substances on the germination of lettuce seeds. Each of four groups of 100 seeds was treated with a different concentration of gibberellin: 0.00 mol dm^{-3}, 0.05 mol dm^{-3}, 0.50 mol dm^{-3} and 5.00 mol dm^{-3}. The seeds in each of the four groups were then subdivided, such that 50 seeds from each gibberellin concentration group were treated with water and the remaining 50 were treated with abscisic acid. All the seeds were then incubated at 25°C under the same conditions, and the percentage germination was recorded.

The results are shown in the following table.

Concentration of gibberellin/mol dm^{-3}	% germination	
	Water	Abscisic acid
0.00	10.0	0.0
0.05	20.0	0.0
0.50	62.0	0.0
5.00	90.0	0.0

Use the results in the table to describe the effect of plant growth substances on the percentage of lettuce seeds that germinate. **4 marks**

....

....

....

....

....

c Why were the seeds kept at a temperature of 25°C? **2 marks**

....

....

....

d A number of variables were controlled during this experiment. Give three variables other than temperature that should be controlled during this investigation. **3 marks**

....

....

....

e Suggest a commercial use for plant growth hormones. **1 mark**

....

....

Topic 6 Animal responses

Animals need to obtain resources from their environment to survive. It is important that animals can respond to environmental changes so that they can increase their chances of obtaining the necessary resources, and so increase their overall likelihood of survival. Animals must be able to detect environmental changes and respond to them in an appropriate, coordinated way, which may involve responses by several different areas of the body. Animals have both a hormonal system and a nervous system to coordinate any responses that they make.

You will need to review the topics on the nervous and hormonal systems.

The human nervous system

There are two parts to the nervous system in all mammals:

- The central nervous system, which consists of the brain and the spinal cord.
- The peripheral nervous system, which is further divided into the somatic and the autonomic nervous system.

1 The brain is complex and its different areas carry out its many different functions. Give details of the functions carried out by each of the five main areas of the brain listed below. (AO1) 5 marks

a **cerebrum**

....

b **hypothalamus**

....

c **pituitary gland**

....

d **cerebellum**

....

e **medulla oblongata**

....

2 Describe the difference between a nerve and a neurone. (AO1) 1 mark

....

Reflexes

3 A reflex action is relatively a simple nervous response, and all reflex actions share specific characteristics.

a **What are the characteristics that are typical of this type of reaction? (AO2)** 2 marks

....

....

b **Give *two* examples of a reflex action. (AO1)** 1 mark

....

4 What is the sequence of five events that occur in any reflex action? (AO2) **3 marks**

....

....

....

5 Name the stimulus and the receptor in the knee jerk response. (AO2) **2 marks**

....

....

6 How is the signal carried from the receptor to the effector in the knee jerk reflex? (AO1, AO2) **4 marks**

....

....

....

7 a What is the survival value of the blinking reflex? (AO1) **2 marks**

....

....

b How does this reflex differ from the knee jerk reflex? (AO2) **2 marks**

....

....

Coordination of the nervous and endocrine systems

8 The nervous system and the endocrine system are both involved in coordination of the animal body. What are the main functional differences between the nervous system and the endocrine system? (AO1, AO2) **4 marks**

....

....

....

9 The flight-or-fight response is an example of a response that involves coordination between both the nervous and endocrine systems. Complete the flow chart to summarise how this response is coordinated. (AO1, AO2) **5 marks**

Environmental danger is detected by sense organs

↓

Information sent to ____________________

↓

Impulse sent to ____________________

↙ ↘

____________________ ____________________

Released by ____________________

10 Adrenaline is an example of a chemical that is involved in cell signalling. What are the main reactions in the body to the hormone adrenaline? (AO1, AO2) **3 marks**

....

....

....

11 The nervous and endocrine systems coordinate in many control mechanisms. One example of this coordination is the control of heart rate. Complete the table below to show the effect each system has on heart rate control. (AO1, AO2) **6 marks**

Nervous system	Endocrine system (....................)

12 What are the main *differences* between the parasympathetic and the sympathetic parts of the autonomic nervous system? (AO2) **4 marks**

....

....

....

....

Muscles and coordination of movement

1 Describe the structures involved in coordinating movement in the body. (AO1) **4 marks**

....

....

....

2 In one type of muscle the structure is made up of bundles of fascicles, which are groups of muscle fibres surrounded by a connective tissue sheath. Each muscle fibre consists of bundles of myofibrils.

a What type of muscle tissue are these structures? (AO1) **1 mark**

....

b The myofibrils (myocytes) are often called muscle cells. In what way are these structures different from other cells? (AO1) **2 marks**

....

....

....

3 **Complete the table to name the equivalent structures found within a muscle cell. (AO1)** 5 marks

Cell structure	Name of structure in a muscle cell
Cell membrane	
Endoplasmic reticulum	
Cytoplasm	
Membrane extensions	
Repeated unit of the myofibril	

The sliding filament model

In this model the myofibrils change in length in order to contract the muscle. They are able to do this because they are made up of two different types of myofilament, known as thick and thin filaments.

4 **What is the name of the protein that forms each type of myofilament? (AO1)** 2 marks

....

....

5 **Label the diagram below of a relaxed myofibril by naming the parts labelled A–I. (AO1)** 4 marks

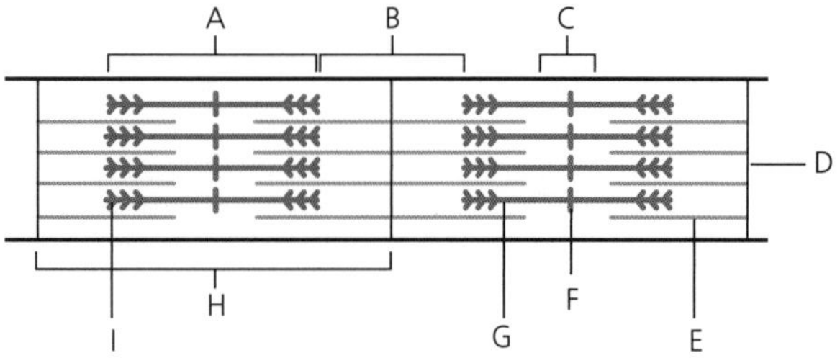

A ....

B ....

C ....

D ....

E ....

F ....

G ....

H ....

I ....

Mechanism of the sliding filament theory

6 **Complete the flow diagram to explain the mechanism involved in the contraction of striated muscles. (AO2)** 8 marks

Nerve impulse ______________

↓

______________ allow depolarisation to spread rapidly through fibre

↓

______________ released from ______________ enter the myofibril

↓

Calcium ions binds to ______________ and shape changes,
so releasing ______________ from binding sites on ______________

↓

______________ binds to binding sites on ______________

↓

______________ released from ______________; the ______________ bends,
pulling the ______________ closer to the centre of the ______________

↓

______________ binds to the ______________ and causes it to release from the ______________;
______________ hydrolyses the ______________ into ______________ and the head extends again

7 **Explain how this process allows the muscle to contract. (AO2)** 3 marks

....

....

....

8 **In an experiment to investigate muscle contraction a few muscle fibres were teased out to give a very fine strand of muscle of about 30 mm long. The strand was carefully straightened on a microscope slide and measured accurately. A few drops of distilled water were added to the slide and after 1 minute the new length was measured and recorded.**

The water was carefully drained before a few drops of glucose solution were added. After 1 minute the new length was again measured and recorded before draining the glucose solution away. Finally, a few drops of ATP solution were added to the slide, and 1 minute later the final measurement was taken and recorded. The results are given in the table.

Solution added	Length measured/mm
None (initial measurement)	31
Distilled water	30.5
Glucose solution	29.5
ATP solution	23

a **Calculate the percentage change in length after each of the three solutions was added. (AO3)** 1 mark b Give two limitations of the procedure as outlined above. (AO2, AO3) **2 marks**

c Describe a potential error in the procedure that will impact on the results. (AO2, AO3) **1 mark**

d There is an error in the results table. What is this error and how might it impact on the results obtained? (AO2, AO3) **2 marks**

e What conclusion is it possible to draw from the results of this experiment? (AO2, AO3) **2 marks**

9 **What is the relationship between ATP and creatine phosphate? (AO2)** **3 marks**

10 **Neuromuscular junctions are specialised synapses. Describe the differences between this type of specialised synapse and other synapses found in the nervous system. (AO1, AO2)** **3 marks**

11 **Complete the sequence of events at a neuromuscular junction. (AO2)** **4 marks**

Impulse arrives at the end of motor neurone, causing the inward flow of calcium ions.

↓

Calcium ions cause ____________________ ____________________

↓

____________________ diffuses across ____________________

↓

____________________ binds to receptors on ____________________ ____________________ ; receptors are ____________________

↓

Calcium channels ____________________ ____________________

↓

The enzyme ____________________ breaks down the neurotransmitter ____________________ ____________________ until the next impulse

Exam-style questions

1 **What causes shortening of the sarcomere during muscle contraction?** 1 mark

A **the actin fibres moving closer to the myosin fibres**

B **calcium ions binding to the actin fibres and causing them to change shape**

C **the myosin head bending and pulling the actin along**

D **the tropomyosin changing shape and pulling on the myosin**

....

2 **In the control of heart rate, would stimulation of the vagus nerve to the heart:** 1 mark

A **increase the output of blood?**

B **increase the rate of heart contractions?**

C **decrease the rate of heart contractions?**

D **stimulate atrial systole?**

....

3 **The hormone adrenaline is an example of a cell signalling molecule that cannot itself pass through the cell membrane of target cells to trigger the release of glucose from the cells. Instead, adrenaline relies upon a mechanism to transmit the information into the cells. Label the diagram below to explain how hormones such as adrenaline exert their influence on the target cell.** 6 marks

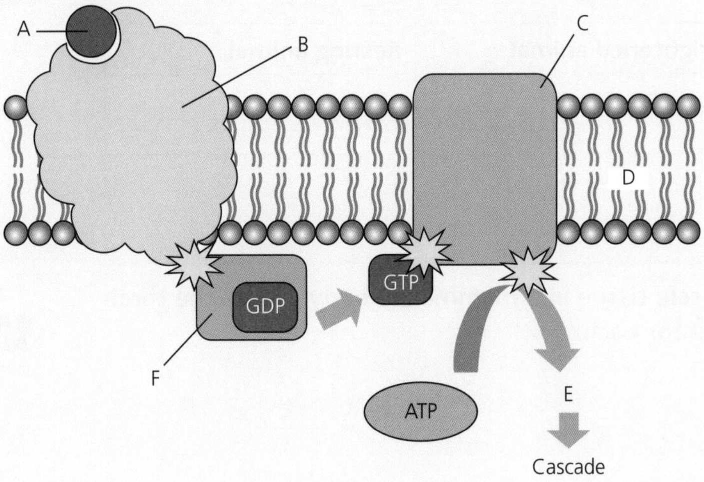

A .... **D** ....

B .... **E** ....

C .... **F** ....

4 The 'flight-or-fight' response results from the combined actions of the hormone adrenaline and the autonomic nervous system, which work together. Some changes associated with this response are external differences in appearance, but there are many changes in how the internal organs function.

a Complete the table to show how two internal organs function differently in an animal that is frightened and in an animal that is resting. **6 marks**

Organ	Organ function in frightened animal	Organ function in resting animal

b The differences given in the table are due to control and coordination by the autonomic nervous system. There are two different aspects to this system, which operate antagonistically to each other. They each use different neurotransmitter molecules to cause their effects in internal organs.

In the table below state the part of the autonomic system that is involved and give the name of the neurotransmitter that will be secreted by neurones into structures such as the heart. **4 marks**

	Frightened animal	Resting animal
Part of the autonomic nervous system activated		
Name of neurotransmitter secreted by neurones		

5 There are three main types of muscle tissue in a mammalian body. Name the three different types and give a function for each. **6 marks**

....

....

....

....

....

6 Briefly describe how the following are involved in contraction of muscle fibres. **3 marks**

a calcium ions

....

b T tubules

....

c phosphocreatine

....

Topic 7 Photosynthesis

Importance of photosynthesis

Photosynthesis is a process that traps light energy and converts it into chemical energy. This is achieved through the use of inorganic molecules such as water and ions, and gases such as carbon dioxide and oxygen. In this way light energy can be converted and used to synthesise food molecules directly for plants, and indirectly for other animals that feed on the plants. Photosynthesis therefore forms the basis of food chains, with all life depending on it, either directly or indirectly, as a food source. It is an essential part of the ecosystem and the food industry.

The processes of photosynthesis and respiration together ensure that the balance of oxygen and carbon dioxide in the atmosphere is kept constant. Aerobic respiration uses oxygen and releases carbon dioxide, whilst photosynthesis uses carbon dioxide and releases oxygen.

1 What name is given to organisms that feed on others organisms? (AO1) 1 mark

....

2 What role do decomposers play in an ecosystem? (AO1) 1 mark

....

Chloroplasts

1 The site of photosynthesis (or photo-autotrophism) in plant cells and some protoctistans is a cell organelle called the chloroplast. Describe the main structures of a chloroplast and state their function. (AO1, AO2) 7 marks

a **chloroplast envelope**

....

....

b **outer membrane**

....

....

c **inner membrane**

....

d **lamellae or thylakoids**

....

e **grana**

....

f **photosystems**

....

g **stroma**

....

2 a How are the pigments arranged and held in the funnel-shaped photosystems? (AO1) 1 mark

b What is the significance of these pigments? (AO2) 2 marks

3 Explain how the two primary pigment reaction centres differ functionally. (AO1, AO2) 3 marks

4 Chromatography is a practical procedure that allows the separation of similar chemicals, such as the photosynthetic pigments.

a Describe the principle that is used to bring about separation. (AO1) 2 marks

b Describe *three* limitations of this procedure. (AO2, AO3) 3 marks

5 A student was provided with a chromatograph and was asked to measure the leading edge of each spot and the solvent front. The results are given below.

Solvent front = 15.0 cm Leading edge: spot X = 12.5 cm spot Y = 11.0 cm spot Z = 9.6 cm

a Calculate the R_f value for each spot, and compare the values with the known values given in the table, to identify one spot that matches one of the pigments. (AO2, AO3) 3 marks

Pigment	R_f value
Carotene	0.95
Xanthophyll	0.71
Chlorophyll a	0.65
Chlorophyll b	0.45

b Give *two* limitations of this procedure that may affect the results. (AO2, AO3) 2 marks

Light-dependent stage

The light-dependent stage refers to the stage of photosynthesis that occurs within the photosystems, in the thylakoid membranes of the chloroplasts. The pigments involved trap and absorb light energy, converting it into chemical energy. This provides the energy required to split water and begins a process that involves transporting electrons through a series of electron carriers. This is similar to the electron transport chain in respiration. As the electrons move through this system their energy is used to produce ATP in a process called chemiosmosis. The products are ATP and reduced NADP, which are used in the light-independent stage of photosynthesis.

1 **What is the most significant feature of the light-dependent stage of photosynthesis? (AO1)** 1 mark

....

2 **Complete the table to show what happens at each of the main light-dependent stages of photosynthesis, and explain their significance in photosynthesis. (AO1, AO2)** 5 marks

Stage of light-dependent reaction	What occurs here	Significance
Light absorption at photosystem II		
Light absorption at photosystem I		
Photolysis		
Electron carriers		
Chemiosmosis		

3 **Explain the term photophosphorylation and briefly describe the essential difference between cyclic and non-cyclic photophosphorylation. (AO1, AO2)** 5 marks

....

....

....

....

....

....

4 **The Z-scheme is a graph showing the energy levels of the electrons on the *y*-axis. The levels increase as sufficient energy is absorbed from the light to raise the electron energy level and then fall as the electrons lose energy to the proton pump. Complete the diagram of the Z scheme below by naming parts A–I. (AO1, AO2)** 5 marks

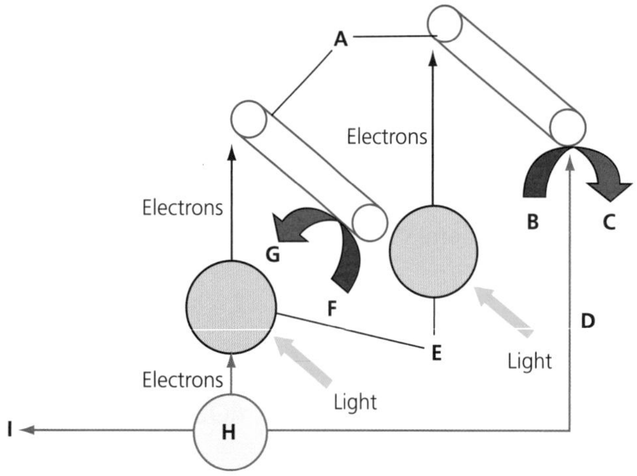

A ....

B ....

C ....

D ....

E ....

F ....

G ....

H ....

I ....

Light-independent stage

The light-independent stage is the stage that occurs in the stroma of the chloroplasts. It uses the products from the light-dependent stage, but does not directly use light energy. However, if light is no longer available this cycle will only continue for as long as the products of the light-dependent stage (ATP and reduced NADP) are available.

The light-independent stage is known as the **Calvin cycle** after Melvin Calvin, who led the team that discovered the biochemistry of the cycle. The carbon source for this cycle is carbon dioxide. It is used to build up complex organic molecules from simple inorganic molecules (carbon dioxide fixation).

1 **Fill in the missing words to complete the paragraph on the light-independent stage of photosynthesis. (AO1)** 10 marks

Carbon dioxide gas enters plant leaves via ____________ ____________.

It then dissolves in water and diffuses into the mesophyll cells and then into the chloroplasts.

***Rubisco* is the enzyme that combines carbon dioxide with a 5C molecule called ____________ ____________ in the stroma. This unstable 6C molecule immediately breaks down into two 3C molecules of ____________ 3-____________ (commonly known as GP) by a process called ____________. GP is said to have become fixed. These two GP molecules are reduced with the addition of hydrogen from the reduced NADP from the light-dependent stage. ATP is used up (creating ADP) during this step, which is called ____________. The result is two molecules of ____________ ____________ (TP).**

To keep the cycle going, five-sixths of the ____________ is used to reform ribulose bisphosphate by further ____________ with ATP. Just one-sixth of the TP forms

_____________ phosphates (phosphorylated sugars), which can be used to form sucrose for plant transport, or to produce _____________ for plant storage.

2 Annotate the diagram of the light-independent stage as a cycle by completing the names and labels. (AO1, AO2) **4 marks**

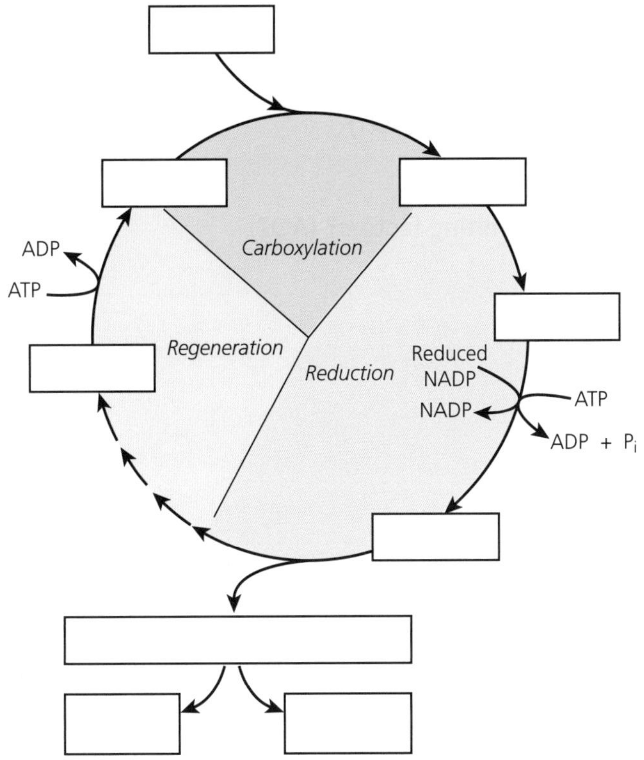

3 Some triose phosphate molecules will leave the chloroplasts for the cytoplasm. How is this possible, given that the chloroplast membrane is impermeable to larger molecules? (AO1) **1 mark**

....

....

4 a Other molecules may be synthesised in the cytoplasm. Name *two* carbohydrate molecules and *three* other biological molecules that may be synthesised. In each case suggest how this synthesis may occur. (AO1, AO2) **5 marks**

....

....

....

....

....

b How many carbon atoms are there in ribulose bisphosphate? (AO1) **1 mark**

....

c What is the enzyme that forms an essential part of the Calvin cycle? (AO1, AO2) **1 mark**

....

Limiting factors

1 **Photosynthesis is said to have limiting factors.**

a **What does this mean? (AO1, AO2)** 2 marks

....

....

b **Name *three* limiting factors that affect photosynthesis. (AO1)** 1 mark

....

c **Why is water not usually included in the list of limiting factors? (AO2)** 3 mark

....

....

....

Temperature

2 **What evidence is there that it is the light-independent stage that is temperature controlled? (AO1, AO2)** 2 mark

....

....

Light intensity

3 **Use the graph below to determine the limiting level of light intensity and explain why it is limiting. (AO1, AO2)** 4 marks

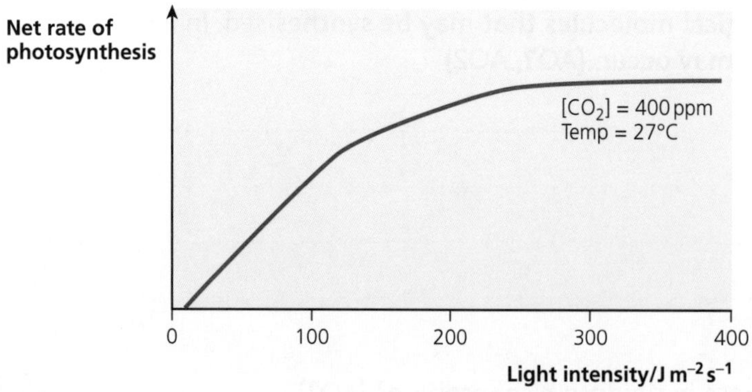

....

....

....

....

Carbon dioxide concentration

4 **Use the graph below to determine the limiting level of carbon dioxide and explain why it is limiting. (AO1, AO2)** 3 marks

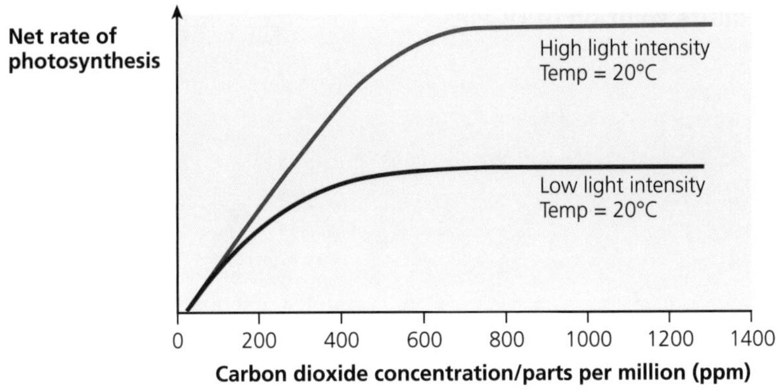

....

....

....

....

....

5 **Why would carbon dioxide become a limiting factor when the level of carbon dioxide in the atmosphere is at an average of 0.04%? (AO1, AO2)** 2 marks

....

....

....

6 **Explain the effect that each of the limiting factors — light intensity, temperature, and carbon dioxide level — has on GP, TP and RuBP. (AO1, AO2)** 3 marks

....

....

....

....

....

17

Exam-style questions

1 **A chloroplast suspension was prepared by grinding nettle leaves in an isotonic buffer solution. This was then filtered through muslin and centrifuged, and the supernatant decanted and thrown away. The chloroplasts were then resuspended in fresh, cold isotonic buffer. A sample of this suspension was used to reduce a dilute solution of DCPiP.**

a **What colour is DCPiP:**

i **when oxidised?** 1 mark

ii **when reduced?** 1 mark

b **What result would you expect in this experiment when using the chloroplast extract prepared and mixed with DCPiP?** 1 mark

c **Explain why:**

i **an isotonic solution was used** 1 mark

ii **the solution was ice cold** 1 mark

iii **the solution was buffered** 1 mark

d **What are the limitations of this procedure?** 2 marks

e **A similar experiment was carried out using a boiled chloroplast extract. What difference would you expect to see in the results?** 1 mark

2 **The optimum temperature for photosynthesis is 25–30°C, depending on the plant species. Between 0°C and 25°C the rate of reaction will double for each 10°C rise in temperature. What is the name of this relationship and what does it tell us about the control of photosynthesis?** 2 marks

....

....

....

....

3 **The metabolic pathway that requires carbon dioxide is an important part of photosynthesis. In what part of the chloroplast does the metabolic pathway that uses carbon dioxide take place?** 1 mark

A **matrix**

B **granum**

C **intergranal lamellae**

D **double membrane around the chloroplast**

4 **The leaves of a plant that is adapted to living in shade are different from the leaves of a plant that is adapted to living in strong sunlight. Suggest *one* way in which the structure of these two types of leaves will differ.** 1 mark

....

....

Hodder Education, an Hachette UK company, Blenheim Court, George Street, Banbury, Oxfordshire OX16 5BH

Orders

Bookpoint Ltd, 130 Milton Drive, Milton Park, Abingdon, Oxfordshire OX14 4SE

tel: 01235 827827

fax: 01235 400401

e-mail: education@bookpoint.co.uk

Lines are open 9.00 a.m.–5.00 p.m., Monday to Saturday, with a 24-hour message answering service. You can also order through www.hoddereducation.co.uk

ISBN 978-1-4718-4731-8

First printed 2016

Impression number 5 4 3 2 1

Year 2020 2019 2018 2017 2016

Printed in Spain

Hachette UK's policy is to use papers that are natural, renewable and recyclable products and made from wood grown in sustainable forests. The logging and manufacturing processes are expected to conform to the environmental regulations of the country of origin.

HODDER EDUCATION
LEARN MORE

ISBN 978-1-4718-4731-8